INTRODUCTION TO POLLEN ANALYSIS

by

R.L. JONES

GEOGRAPHY DEPARTMENT

LANCHESTER POLYTECHNIC, COVENTRY

AND

P.R. CUNDILL

GEOGRAPHY DEPARTMENT

UNIVERSITY OF ST. ANDREWS

TABLE OF CONTENTS

EDITORS' PREFACE

The publication of the Technical Bulletins by the British Geomorphological Research Group arises out of a common need, expressed by the Group as a whole, to have a source of standardised information relating to increasingly sophisticated methods of data collection and data handling. The demand for such Bulletins is three-fold: for comparability of results among workers in the field; for the dissemination of information concerning techniques, particularly where this speeds up and facilitates the execution of a research programme, and for manuals which stress particularly the uses and limitations of the techniques in the field of geomorphology.

The authors and editor have endeavoured to ensure accuracy and where necessary the manuscripts are being forwarded before publication to interested commercial firms for critical evaluation. In other respects the authors alone are responsible for the views expressed.

EDITOR:-

Dr. T.C. Atkinson, School of Environmental Sciences,
University of East Anglia,
Norwich, NR4 7TJ,
England.

INTRODUCTION

Pollen analysis of Quaternary deposits is well established as a major technique in the reconstruction of past environments and this has numerous possible applications in geomorphology. It should be recognised from the outset that pollen analysis is not the only method of examining sediments and while this paper concentrates on one technique other techniques such as macro-fossil analysis (e.g. Bell 1970, Watts and Winter 1966), diatom analysis (e.g. Evans 1970, Haworth 1969) and chemical analysis (e.g. Mackereth 1966) will yield valuable results. The use of the term 'pollen' analysis may be misleading for it is intended that it should include the isolation and identification of spores of Pteridophyta and Bryophyta as well as pollen of Angiospermae and Gymnospermae.

Although the recognition of pollen grains and spores as air-borne particles and microfossils in rocks dates back well in the nineteenth century, it was not until the early twentieth century that pollen was extracted from organic Quaternary deposits, the value for each different pollen type expressed in percentage terms and then interpreted as past vegetation patterns. Lennart von Post during the second decade of the century was one of the earliest research workers to utilize the technique and report its potentialities to a wide scientific audience (von Post 1918). During the 1920s knowledge of pollen analysis spread throughout Europe, particularly through the work of Erdtman (1928, 1929). In Britain, early protagonists of the technique included Blackburn (1931) and Godwin (1934). More recently, the use of pollen analysis has spread to many parts of the world and has led to a very great increase in detailed knowledge of past environments. (Bakker 1964, van der Hammen and Gonsales 1965, Cushing 1967, Fredskild 1969, Hope 1976).

Pollen analysis relies on the fact that all flowering plants produce pollen and ferns and mosses produce spores as part of the process of reproduction. Pollen and spores from different species can be identified by their different characteristics. A resistant outer casing allows pollen and spores to be preserved in sediments although this casing can be destroyed by oxidation processes. Pollen and spores may normally be preserved in three major sites: lakes, peat bogs and soils.

The study of fossil pollen is being used increasingly to elucidate problems of past distributions of plants and palaeo-environments within historical geography, biogeography, pedology, climatology and geomorphology. A number of areas within geo-morphology have become well documented through the use of the technique with perhaps the major application in the reconstruc-tion of interglacial and interstadial environments (West 1957, Watts

1959, Turner 1970, Birks 1973), as a result of which a much better understanding of individual interglacial and interstadial periods has been achieved. Pollen analysis has been linked with geomorphological evidence in the reconstruction of the pattern of land and sea level changes which occurred in some lowland areas of Britain (Godwin 1943, Newey 1966, Brooks 1972 and Tooley 1976). The technique has also been used in attempts to examine the history of landslips (Franks and Johnson 1964, Simmons and Cundill 1974) and has provided information from palaeosols about geomorphologically and pedologically interesting situations (Andersen 1965, Paepe 1971). There are many possible geomorphological situations where if organic material exists it could be analysed for pollen and subsequently used as an additional line of evidence.

There are a number of texts or articles describing the techniques of pollen analysis, for example, Brown (1960), Kummel and Raup (1965), West (1968), Faegri (1975) and Barber (1976). It is the intention of this bulletin to present the simplest methods, using the minimum amount of equipment in order to obtain satisfactory results from the technique. It is not, nor could hope to be, a full discussion of all aspects of pollen analysis. However, it is hoped that it may encourage more research workers to sample and analyse potentially polleniferous deposits. Advice on the more difficult aspect of the interpretation of results could be sought from experienced palynologists.

STRATIGRAPHICAL INVESTIGATIONS

Introduction

It is necessary to acquire some knowledge of the matrix from which pollen grains are hopefully to be recovered in order to evaluate the results. The recording of stratigraphy will be a process familiar to many geomorphologists and the examination of biogenic sediments should be carried out in the same critical way as that of minerogenic deposits.

A number of factors are important in this context. Firstly, plants which formed a deposit have a pollen production of their own (a local production) and this will need if possible to be isolated from the more representative regional pollen assemblage reaching the site. Secondly, the physical and chemical characteristics of a deposit will influence the state of preservation of pollen grains and spores. Knowledge of this may at the outset suggest that palynological investigation may be unprofitable and it will certainly indicate which laboratory treatment to use for optimum results. Thirdly, an understanding of plant communities forming sediments such as peat can be useful in the interpretation of palaeoclimates.

Sediments can be of two main types; autochthonous (e.g. peat) consisting of remains of vegetation once living in a particular place from which a sample is recovered, and allochthonous (e.g. mud) consisting of material which originated elsewhere and has been deposited or redeposited in the place of sample recovery.

The primary aims of palaeobotanical fieldwork include the collection of uncontaminated samples of the deposit, the determination and recording of the types of sediment and the explanation of stratigraphical relationships. Often there is no choice of locality, as, for example in temporary exposures such as pipelines or motorway cuttings, but if an integrated synthesis of the development of the terrestrial vegetation of an area is required certain prerequisites are necessary. Ideally small basins containing lakes or peat bogs should be avoided because of their tendency to reflect local conditions in their pollen record. Any type of peat bog has pollen produced locally by herbs and small shrubs. Lake deposits are preferable - local aquatic and reedswamp pollens are easier to isolate and grains generally are well preserved. However, questions of erosional hiatuses by wave action or changes in lake level and redeposition of sediments by seasonal overturn frequently occur in such situations.

Sampling from open sections

In any event there are a number of cardinal rules which must be observed in the examination and collection of sediment samples. Whatever the type of deposit, an open section is most satisfactory for analysis. Sediments tend to vary laterally very considerably over short distances and the most complete sequence thereabouts can be delimited and chosen for sampling. After cleaning the face a section should be recorded using some appropriate notation for each sediment type (Fig.1), drawn to a convenient scale and photographed. The column selected for sampling should be cleaned by horizontal cuts of a sharp knife so as not to carry material from one layer to another. A tape measure is then stretched vertically down the section and either sample tubes (5 x 2cm) pressed into its wall or small pieces of sediment (about 5cc is enough for pollen analysis) removed by broad smooth forceps of a spatula. A 5cm sampling interval is most common, with additional samples taken at sediment contacts, but compressed sections require a smaller interval perhaps of 2cm between samples. Each sample bottle is labelled with a site name and its depth, corked and stored for removal to the laboratory. The sampling instruments must be cleaned thoroughly between each sample - never by a piece of cloth which is used again but by some disposable material (e.g. tissues).

For more exact work and for such purposes as macrofossil analysis and radiocarbon assay a monolith should be removed from the section. Convenient dimensions for a monolith are 15cm wide and 15cm deep by 50cm long. Construction of the box should be of strong metal (e.g. stainless steel - although this is expensive) which can be hammered into sediments after lining it with poly-

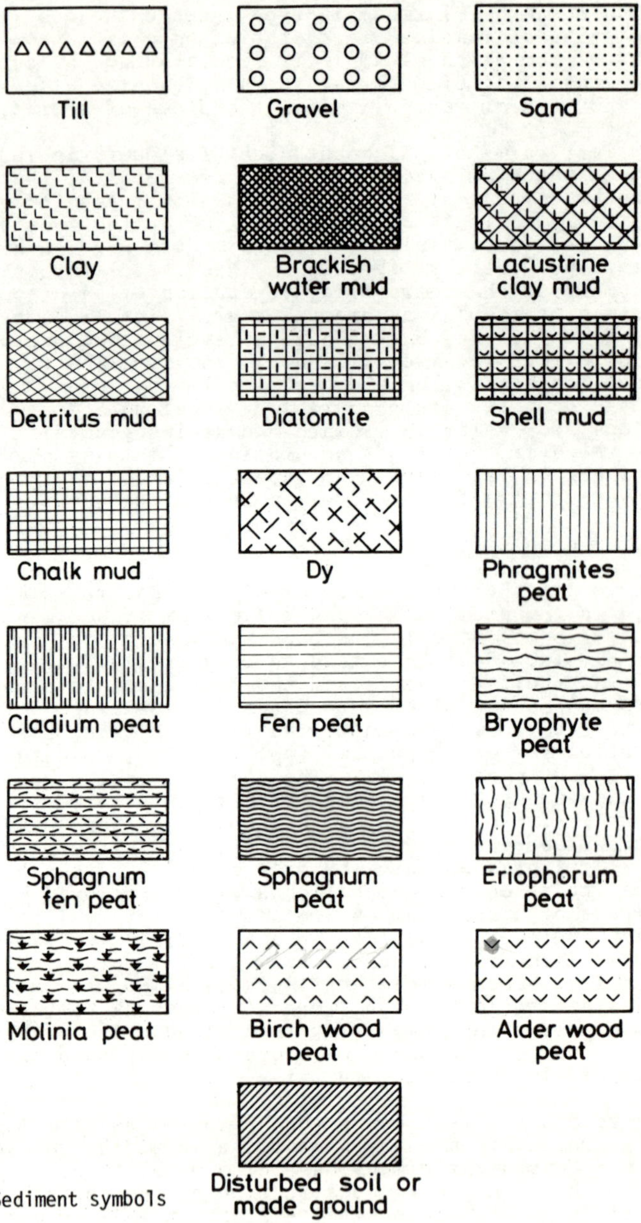

Fig. 1 Sediment symbols

thene and labelling it appropriately. Removal is effected by cutting back into the surrounding sediment at a sufficient distance from the monolith, prizing out the section which is then squared off, wrapped in polythene and transported to the laboratory.

Sediment samplers

Very often there is no alternative but to investigate stratigraphy and procure samples with some form of coring instrument. The appropriate tool will depend upon the sediment, the depth to which it is necessary to penetrate, the manpower available, the ease of access to the site and the cost of such an operation. According to West (1968) the key to successful coring is ingenuity in adapting equipment to field conditions and in dealing with unexpected problems. Frequent damage to equipment occurs (e.g. sub-surface impaction, buckling or total loss of instrument) and a basic tool kit, toghether with knowledge of methods such as that of Clymo (1964) for the recovery of lost peat borers is necessary.

Most basic samplers are made commercially in Britain or can be obtained through suppliers. Alternatively, plans which are not beyond the scope of institutional workshops to execute can be sought from designers or published sources. A list of suppliers is given in Appendix 2. The general principle of a sampler is that of a closed chamber of stainless steel attached to a rigid string of alloy or steel rods being pushed vertically down through a deposit to the required depth where it is opened and filled either through its side or bottom. Whenever possible all instruments should be pushed rather than screwed down as turning damages stratigraphy and hence sampling potential. The rods are coupled by set screws in male/female fashion or by spring loaded pins with a coupling enclosed by a sliding tube. Lengths of rod vary from ½ to 1½ metres and ½m intervals are usually indicated where necessary along them.

The Hiller sampler

Hand driven samplers are most commonly employed and the most frequently encountered of these is some type of Hiller borer (Fig.2). This consists of a 30-50cm long sampling chamber 2-3cm in diameter with a 2cm slit down one side. The chamber is covered by an outer tube which can move around it and has a similar sized slit bordered by a 2-3cm wide flange set at a tangent and with a sharpened edge (Fig.2). One model (Thomas 1964) has inner removable liners of a light alloy for the sampling chamber so that a complete core can be extricated for laboratory examination. The mechanics of sampling involve the insertion of the chamber in a closed position using slight clockwise pressure so that the outer tube remains covering the slit of the inner tube. When the desired depth is reached, anticlockwise rotation opens the instrument by making the two slits coincide. Two or three further anticlockwise turns force the flange to scoop sediment into the inner tube. Clockwise rotation closes off the inner tube

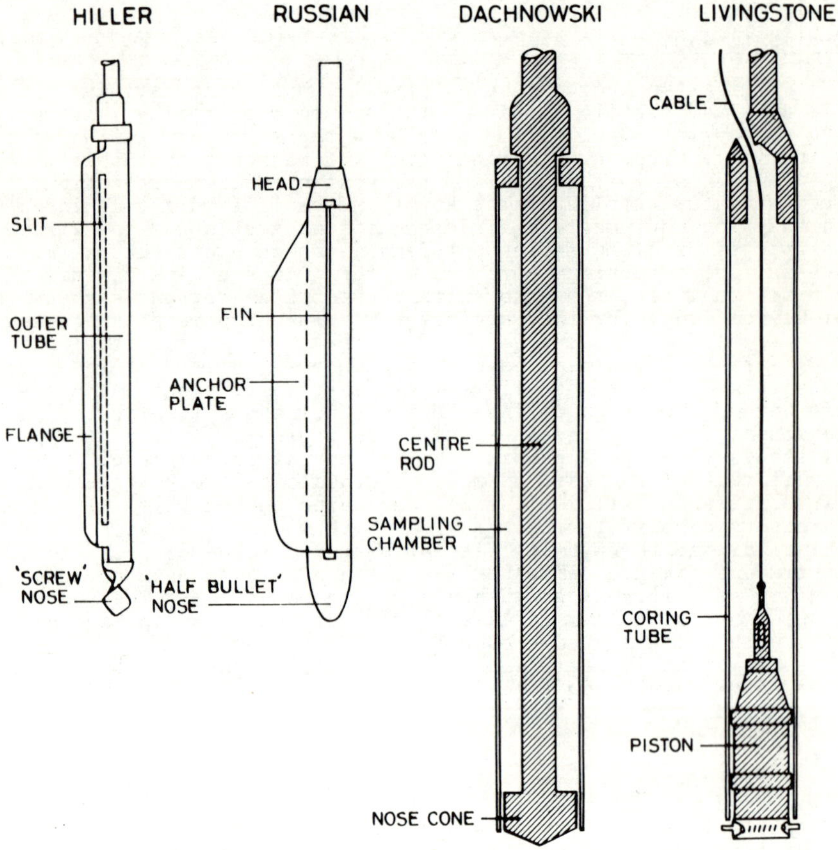

Fig. 2 Coring instruments

and withdrawal using slight clockwise pressure ensures that the chamber remains shut. If a liner is not inserted, samples are removed from the centre of the chamber using the same principles adopted for sampling open sections. The Hiller sampler is reasonably portable and fairly easy to use. It does not compress sediments and takes acceptable samples in most peats and muds. It distorts the sediment as the flange forces it into the inner tube and displaces stratigraphy as it is inserted. This and other hand samplers are frequently operated in alternate boreholes a few cm apart so as to pass through an uncontaminated length of deposit before each new core is taken, though the dangers of lateral variations even in such small distances have been apparent to many workers. The Hiller sampler is also jammed by sand and stiff stony clays. It will not hold loose or wet sediment which during ascent or descent may either leave, or more seriously, enter the chamber. The chamber, especially in the smaller models, in difficult to clean properly - if water is not available on site a supply should be taken in a plastic container for this purpose. together with a garden syringe.

The Russian sampler

The 'Russian' type sampler (Fig.2.) became 'The Peat Survey of Scotland' sampler when Jowsey (1966) described in detail its construction. Other workers e.g. Barber (1976), have modified the basic design but the principle of operation remains unaltered. A half-cylinder of stainless steel 50cm long and 5 cm in diameter fixed to a sampler head is rotated at depth through 180° so that a half cylinder of sediment is retained against a central anchor plate (Fig.2). Originally the sampler was meant to take two quarter cylinders of sediment (one for field examination, one for laboratory analysis). To accomplish this, one side of the anchor plate was finned and this was on the outside when the chamber was inserted. The full half core is obtained by the opposite procedure, with the obvious disadvantage of sediment disturbance by the fin avoided and provision also made for the removal of the core in half drainpipes (5cm diameter and 50cm in length) made of strong plastic. These are labelled with successive depth measurements and either wrapped in aluminium foil or polythene sleeving of suitable dimension. This type of sampler preserves sediment structures, is easily cleaned but must not be turned while being inserted, reducing its effectiveness in wood peats or mineral horizons. Such deposits can be penetrated only through a series of jerk motions brought about by the pressure of 2 to 3 individuals on the handle. Jamming of pebbles or wood between the anchor plate and chamber - partially overcome by careful turning - can cause serious twisting of the sampler above the nose.

The Dachnowski sampler

This corer is an end-filler to which recent modifications have been made by Walker and Lowe (1976) who retitle the instrument the Abbey Corer (Fig.2.). As in other types the basic operational principles are similar to the original model with a retractable piston attached to the end of a rod bearing a 60cm long and 5cm diameter tube made of metal (Fig.2). The sampler is lowered with the piston positioned at the end of the tube and when the sampling depth is reached the piston is raised to the top of the tube by withdrawing the rods until it is held here by a spring catch on the basal rod engaging a collar, or by giving the rods a half-turn and engaging splines cut in this rod and working in grooves cut in the tube against the top of the tube. The sampler can then be pushed down to a depth corresponding to the length of the tube and fills with a core. The sampler is raised and the core extruded on to half plastic drainpipe by pushing back the piston. The design is fairly simple and the instrument not unduly heavy to transport though the 'Abbey' model can be criticised on both these grounds. All types give reasonable penetration in fine grained, stiff sediments and take undisturbed cores but they do not retain loose deposits and are subject to contamination by leakage through the chamber top. Furthermore, compression is liable to occur in the sediment.

The Livingstone sampler

Stationary piston corers of the Livingstone type are widely used at present (Fig.2). Here the piston remains stationary while the coring tube is driven downwards. A typical model (Fig.2) consists of a removable tube of light alloy about 1 metre long and 10cm wide fixed to a head by screws. The piston consists of 2 rubber stoppers on a threaded bolt held in the base of the tube by spring-loaded pins and is attached to a wire running to the ground surface. The sampler is operated by pushing to a required depth when the wire is fixed with a lever grip. The sampler is then driven down for the length of the tube past the stationary piston. The sampler is filled, raised and the core extruded as described for the Dachnowski, or the tube is removed, corked at the ends and a new one fitted for the next sample. Hence, cores can be removed in rapid succession, with little deformation of the sediment and of a suitable (10cm) diameter for thin horizontal slices to be taken for radiocarbon assay. Its disadvantages include compression and a reluctance to sample coarse or loose material.

The Mackereth sampler

In addition to the hand operated samplers there is a more sophisticated and more expensive sampler which can be used to collect cores from lake sediments. This is the Mackereth sampler (Mackereth 1958) which operates through compressed air action and can be used only where there is reasonable water depth (e.g. in excess of 4m for a 6m sampler).

Soil augers

Finally mention should be made as to the usefulness of various types of auger as sampling tools for sediments. They are not ideal, giving a distorted and often contaminated sample by the time they reach the surface, but are a valuable aid, not least in the penetration of initial dry or coarse deposits in the upper part of boreholes, or with extension rods, similar layers at depth before conventional sampling can continue. Of the many types available the bucket and post-hole (Jarrett) varieties seem to yield the best results.

Storage of samples

Ideally the collected material should be examined as quickly as possible. Samples brought from the field in tubes which are corked or in drainpipes covered with foil or polythene will be suitable for analysis in this form for a few months. After this time desiccation and fungal growth will begin, so if long storage is suspected, tubes should sealed with paraffin wax and cores frozen. In all instances, storage for any length of time should be in a low-temperature, damp environment where chemical changes and drying out will be minimised.

SETTING UP A POLLEN LABORATORY

An individual embarking upon the design and equipping of a laboratory suitable for exacting pollen analysis (though it must be stressed that more or less adequate results may be obtained employing less rigorous standards) will encounter both complexity and expense. Depending upon whether a research or teaching function (or both) is intended the amount of space required is difficult to specify. If one person is to occupy a basic laboratory, quite small dimensions (about 10 x 4 m) will suffice; for teaching at least ten times as much space may be necessary. Costing is even more difficult. Most items are expensive. Actual sources of supply are easier and Appendix 2 forms a consolidated list of those found both reliable and economical, though no preference is implied.

Ideally, the laboratory should be sterile with windows that do not open and an air extraction system, as well as a double entry door separated by a short length of corridor served by another extractor fan to remove atmospheric contaminants. Alternatively a glass fibre filter system on the doors will also help reduce sources of possible contamination. Essential is a fume cupboard of large dimensions and sturdy construction (especially of anticorrosive glass) containing a sink (not lead lined) and a motor capable of drawing up fumes of noxious and heavy chemicals such as hydrofluoric acid and acetone. Also vital is a centrifuge which must be capable of safe operations at speeds of at least 3000 rpm and which has interchangeable swing-out heads for 8, 15 and 50 ml tubes. A range of smaller items are also absolutely necessary. This includes polypropylene test tubes and funnels, caps, beakers, reagent bottles and petri-dishes, together with high quality small diameter sieves of a mesh fine enough to allow the passage of pollen grains and spores but not extraneous material (about 100-150 mesh to the inch). These can be used until damaged, when they must be replaced to avoid risk of accidents or contamination. Consumable items include a range of chemicals such as sodium hydroxide and hydrochloric acid, sample storage bottles, cotton wool, filter paper and staining material, together with microscope slides and coverslips.

Detailed requirements for a microscope are described on page 18. The recording of pollen and spores is usually carried out on standardised sheets containing the names of commonly encountered taxa, although to speed up the process use can be made of mechanical or electrical counting instruments. Laboratory manuals are essential if pollen analysis is undertaken as a serious exercise and should include:

Textbook of Pollen Analysis	Faegri (1975) (3rd Edition)
Handbook of Paleontological Techniques	Kummel and Raup (Eds) (1965)
Palynological Techniques	Brown (1960)
Handbook of Palynology	Erdtman (1969)

together with the relevant parts of works such as

Methods in Plant Ecology	Chapman (Ed) (1976)
Pleistocene Geology and Biology	West (1968)

THE PREPARATION OF SAMPLES FOR POLLEN ANALYSIS

Introduction

There are two methods of analysing pollen: firstly by use of relative pollen counts and secondly by employing absolute pollen frequencies. The former is more traditional and widely used, and although the preparation technique is simpler and less time consuming than absolute analysis, there are more complications in the interpretation of the resulting data. Both methods are outlined below with the relative method described first.

The ideal of a pollen preparation is to produce a suspension of fossil pollen with as little matrix of extraneous material (plant debris and silica) as possible in order to obtain a clear view of the grains. Most sediments present some special problem requiring a prescribed treatment perhaps involving several chemical reactions. However, as Barber (1976) stresses, the indiscriminate application of successive techniques of preparation in order to try and achieve a pure suspension will almost certainly lead to disaster in the guise of differential destruction and collapse of the pollen grains.

Most techniques have a common basis though modifications are made from worker to worker through time in the light of experience and advances in methodology. The sequential account that follows is a compendium of methods employed over the last decade at the Universities of Durham, Lancaster and Cambridge and to the many colleagues who have contributed suggestions we owe acknowledgement. The aim is to use effective and safe methods to remove carbonates, humic colloids, silica and cellulose. A small amount of sediment is placed in a boiling tube using a clean spatula and the relevant chemical added. Samples are normally prepared in batches of 4 - 12 (depending on the centrifuge head capacity) and all pieces of equipment which will come into contact with individual samples are assigned a sequential number in order to avoid misuse and confusion. A record of sample numbers, depths and notes on any problems associated with particular samples should be kept in a laboratory notebook. Heating is often necessary and this is followed by sieving to remove as much of the non-polleniferous matrix as possible. Macrofossils of over sieve mesh size remain for supplementary examination at this stage. Washing, centrifugation, staining and mounting of the polleniferous residue follows. If a number of processes are being employed it is useful to prepare temporary slide mounts after each in order to check progress.

Certain steps in all preparations are enacted in the same way using standard equipment and to avoid repetition they may be specified. Samples in 50 or 15ml polypropylene centrifuge tubes are balanced with an equal amount of liquid in each tube and centrifuged at a speed of 3000 rpm for 5 minutes (for 50ml tubes) and 3000 rpm for 2 minutes (for 15ml tubes). After centrifugation the supernatant liquid is carefully poured off in one movement.

The procedures are outlined below in sequential order assuming the complete set of treatments are applied to an initial sample. Note that % figures are based on volumetric measurements. When making up dilutions of acid always add acid to water.

A. Carbonate Solution and Alkali Digestion of Humic Compounds

1. Place *circa* 2gm of sediment in a 50ml polypropylene boiling tube and add a little 10% HCl. Fill to 2/3 if reaction is gentle.

2. When the reaction is complete, centrifuge and decant.

3. Add a few drops of 10% NaOH or 5% KOH to the residue in the tube, mix then add a further 20ml and place in a boiling water bath in a fume cupboard long enough for the thorough breakdown (no lumps remaining) of the sample (10 min - 1hr).

4. Filter the mixture through a sieve into a polypropylene centrifuge tube. Wash the residue on the sieve thoroughly with distilled water supplied by a jet, and direct the water into the centrifuge tube.

5. Centrifuge and decant. If the supernatant liquid is dark, substantial amounts of humic material will still be present in the sample.

6. Wash the sample in the tube with distilled water, and centrifuge and decant. Repeat this process until the supernatant liquid becomes clear.

7. Wash the residue on the sieve towards its centre with a jet of water. Invert the sieve and wash the residue into a petri-dish for macrofossil examination. Thoroughly wash sieves after use to avoid contaminating later samples.

For many organic deposits the procedures in A will have removed adequately most of the plant debris and concentrated the pollen, and in such circumstances proceed to section D. Samples which contain large amounts of silica and/or plant debris should be treated as in sections B and C before going on to D. Rubber gloves and eye protectors must accompany a laboratory coat when using concentrated chemicals in procedures B and C. Concentrated acids must only be used in an operational fume cupboard which ought to be checked by a Safety Officer to ensure that it is of adequate design and performance for use with such acids. The laboratory must contain a first aid kit equipped with (amongst standard items) a syringe and a solution of 10% calcium gluconate for injection under the coagulum of hydrofluoric acid burns. The dangers of using hydrofluoric acid mean that this part of the technique should not be undertaken in teaching undergraduate classes.

<u>B. Removal of Silica using Hydrofluoric Acid</u>

8. Add a few drops of distilled water to the residue in the polypropylene centrifuge tube and mix thoroughly.

9. Add 20 ml of 40-60% HF to the tube and place in a boiling water bath (or a sand bath) for as long as is necessary (between 20 and 60 mins is usually sufficient) to eradicate most of the silica (test for residual mineral particles with a clean stirrer).

10. Balance with 10% HCl, centrifuge (capped centrifuge tubes are preferable) and decant carefully into the fume cupboard sink while this is being flushed continuously with water.

11. Add a few drops of 10% HCl, mix, then a further 20 ml of 10% HCl and place in a boiling water bath for 15 minutes to remove colloidal silicon dioxide and silicofluorides.

12. Centrifuge and decant.

13. Add a few drops of 10% NaOH to the next washing water, centrifuge and decant. Repeat using distilled water only.

C. Acetylation to remove Cellulose

14. Add 10 ml glacial acetic acid to residue in tube, mix and centrifuge. Decant into fume cupboard sink being flushed continuously with water. Repeat.

15. Prepare acetylation mixture by adding 1 ml concentrated H_2SO_4 to 9 ml acetic anhydride. Mix thoroughly in a measuring cylinder with a glass rod in order to prevent an exothermic reaction.

16. Add a few drops to the sample, mix and then fill the tube with 20 ml of mixture. Put in a boiling water bath for 3 min.

17. Balance with acetylation mixture, centrifuge and decant.

18. Add glacial acetic acid, mix, centrifuge, and decant.

19. Add distilled water containing a few drops of 10% NaOH, mix, centrifuge and decant. Repeat using distilled water only.

D. Preparation of Pollen Mounts

Either (a) Temporary. Here pollen grains can be moved around upon the slide but dehydration of the mount occurs quite quickly.

20. To residue in the tube add a few drops of distilled water, 1 or 2 drops of safranin or fuschin stain, mix, centrifuge and decant.

21. Invert tube carefully over a filter paper and allow to drain.

22. Add 3 - 6 drops of glycerol and mix thoroughly with the residue. The amount of glycerol required will depend upon the amount of residue and the pollen concentration within it. Test this by preparing a trial slide.

23. Using a small spatula transfer a drop of material to a clean glass slide of 1.0 mm thickness and place a 22 x 40 mm cover slip (No. 1½) over it. Label the slide with a sample number, the site notation and its depth in the profile.

24. Seal the cover slip with nail polish taking care to keep pollen material off the varnish brush.

25. Transfer remainder of preparation to a small appropriately labelled glass specimen tube with a polythene stopper. Store for future use.

Or (b) Permanent. Here a liquid medium is used which has high viscosity and therefore does not allow grains to be moved easily on the slide. However, it does not dry out or alter grain size much.

20. Add distilled water, mix, centrifuge and decant.

21. Add 1 ml distilled water, 5 ml 100% ethanol, centrifuge and decant.

22. Add 100% ethanol, mix, centrifuge and decant.

23. Add *circa* 1 ml of toluene and pour mixture from the centrifuge tube into a labelled glass tube sealed with a polythene top. Balance with Toluene.

24. Carefully lower tube into centrifuge using forceps. Centrifuge for 10 minutes at 750 rpm.

25. Decant into a beaker in the fume cupboard.

26. Add 2 - 6 drops of silicone fluid (200/2000 centistokes viscosity). Mix well to avoid clumping of the pollen grains in the preparation.

27. Allow excess toluene to evaporate for 24 hours in the fume cupboard.

28. Make up a slide fixing the coverslip with a blob of varnish at each corner - there is no need to seal completely.

Dimbleby (1961) describes a method of extraction of pollen from soils, but as this is based on preparation techniques outlined by Faegri (1975) and differs only in detail from the techniques in this paper, it is not included as a separate description.

A variety of other techniques of extraction exist, such as silica by differential flotation and lignin by oxidation. For further information, the most useful source of reference is *Handbook of Paleontological Techniques* (Eds Kummel and Raup 1965), now unfortunately out of print.

Absolute Pollen Analysis

When pollen has been prepared using the schedules already described the normal procedure is to express the number of grains of one taxon as a percentage of some total number of grains counted in the sample - the so called 'relative' method of pollen analysis. Absolute pollen analysis requires either the introduction of a known quantity of exotic pollen or spores into a measured quantity of sediment sample and fossil grains counted are related to the total count of the introduced type, or the sampling of a known proportion of the total population of pollen by taking a known weight or volume of sample for counting (Peck 1974). The absolute method allows for easier interpretation of results as percentages of fossil pollen grains are independent of each other. Comparisons of pollen diagrams using both absolute and relative methods on indentical sediment samples have revealed marked differences in some instances. This is especially true in time zones where it is known that substantial increases or decreases in pollen production occurred (the late-glacial/post-glacial boundary at the start of an interglacial period for example). When radiocarbon dates are available the rate of accumulation of sediment can be combined with a measure of pollen concentration within it giving a figure for pollen influx (expressed in grains cm^{-2} $year^{-1}$).

The procedures involved in absolute pollen analysis are lengthy and complex with many workers adapting and comparing methods. Some of the complexity has been reduced by the introduction of 'pollen pills' (Stockmarr (1972), Appendix 2) which are assayed concentrations of exotic pollen or spore types in a convenient form. The most important papers on absolute pollen analysis include Beninghoff (1962), Davis (1965a, 1967), Matthews (1969), Waddington (1969), Pennington and Bonny (1970), Bonny (1972), Pennington (1973) and Peck (1974).

In outline *one* method of absolute pollen analysis may be summarised thus;

1. Remove adjacent cm samples of approximately 3 cm^3 from the large (e.g. 10 cm) diameter sediment core carefully using a cork borer.

2. Extrude these into airtight bottles and store at a low temperature (e.g. 5°C) until required. The remainder of the core may be used for radiocarbon assays.

3. From the sample in the bottle measure out 0.5 cm^3 of sediment by displacement in water in a narrow bore 10 cm^3 graduated cylinder (use the same cylinder for every sample and read menuscus change between same levels every time).

4. Add a pollen pill of exotic pollen or spores to the sample.

5. Prepare the sample using the treatments described for the relative method.

6. Mount the sample as described for the relative method.

7. Make up slides using aliquots.

8. Count all pollen (including exotics) along regularly spaced traverses until a total of 500 exotics has been reached.

9. Calculate the pollen concentration by employing the appropriate formulae.

10. If radiocarbon dates are available calculate the pollen influx figures by dividing the pollen concentration by the

rate of sediment accumulation as determined by successive radiocarbon assays. This will be x cm C^{14} $year^{-1}$ and the final value will be expressed as grains cm^{-2} C^{14} $year^{-1}$

Absolute analyses require considerable extra preparation, counting and calculation and, for a time based method, sufficient radiocarbon dates. It is likely that relative methods will remain useful for routine investigations; absolute counting being reserved for special problems.

POLLEN IDENTIFICATION AND COUNTING

The microscope and pollen counting

Once chemical preparation has been completed detailed examination of the residual material is possible. Because of the small size of pollen and spores (usually between 15 µm and 50 µm) it is necessary to use a high-powered microscope in order to be able to identify them. There are a number of minimum requirements for the microscope in use. It must have a series of obnectives (e.g. x 10, x 40, and usually an oil immersion lens of x 100) which when combined with eyepiece magnification of x 10 will give total magnification values of e.g. x 100, x 400 and x 1000.

A coaxial mechanical stage and binocular head are essential for pollen counting (which is simply the methodical traversing of the slide identifying and noting all pollen types encountered). Double-counting will be avoided by ensuring that there are at least 1½ fields of view between each traverse of the slide. Brookes and Thomas (1967) and Peck (1974) discuss the non-randomness of pollen grain distribution on microscope slides while Crabtree (1975) describes work which showed no significant difference between counts made on part of a slide compared with a whole slide (see also Peck 1974). The total number of pollen grains counted should be statistically valid. This is discussed in the section on Pollen Statistics and the Pollen Diagram.

General identification and counting of pollen and spores should be carried out at a magnification of between x 300 and x 600 as this will be sufficent to clarify differences between most of the more common pollen and spore types. However, certain detailed features of surface structure can be seen only under high magnification, and this is when use is made of the oil immersion objective, giving magnification of x 1000 or more. In other situations surface structure details can be examined if the microscope is also fitted with phase contrast equipment.

Pollen identification

The major problem when starting pollen analysis is the identification of individual pollen grains. The best advice is to look at, in the first instance, photomicrographs of pollen and

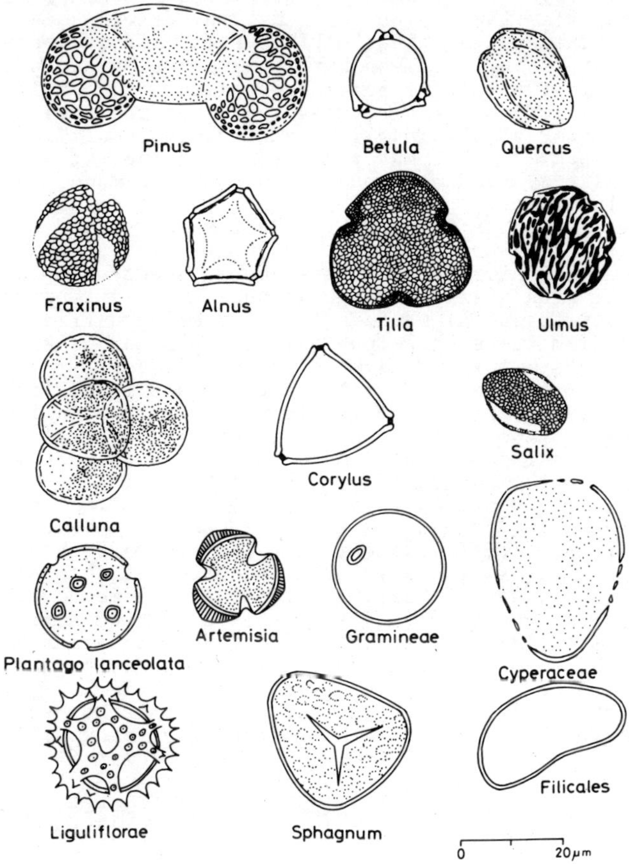

Fig. 3 Selected pollen grains and spores

spores. General photomicrographs of pollen and spore material as it appears down a microscope are given in, for example, West (1968, 1971) and Sparks and West (1972), and more detailed photomicrographs of individual pollen and spore types are given in pollen atlases (Hyde and Adams 1958, Erdtman, Berglund and Praglowski 1961). Drawings of pollen and spores (Fig.3) provide a subsititute for photomicrographs. Major types are illustrated in Gregory (1973) and Godwin (1975), while Erdtman's volume on pollen analysis (1943) provides many line drawings in its illustrations. These suggestions may be useful for making a reasonable start to identification but it is normally accepted by palynologists that a type or reference collection of pollen material is essential for complete accuracy. A type collection is pollen and spore material obtained from accurately identified present day plant species. The material from each individual species is chemically treated and mounted in the same manner as fossil material in order to ensure comparability between the two (Appendix 3). In using a type collection, a second microscope for comparison is useful and although it need not be so refined as the counting microscope it must be as powerful. For example, a monocular with a simple stage is adequate. A type collection is obviously essential when analysis is being carried out on sediments from areas where photomicrographs and drawings of the pollen flora are not readily available. In some areas the identification of pollen has developed to such a sophisticated level that keys have been devised to aid preliminary identification of the pollen type. One such key is provided by Faegri (1975), based on the structure and ornamentation that is apparent on the surface of pollen and spores. Keys of this kind require a knowledge of fairly detailed terminology and if pollen analysis is being carried out regularly it is worthwhile to become familiar with their use. If infrequent analyses are made then the time spent trying to use the key would be out of all proportion to the value of the results. One further point to remember when identifying pollen and spores is that they are three-dimensional and because processing removes the inner tissue they may be easily folded. This may mean that they are in a position which is not the one depicted in photomicrographs and drawings. A very gentle touch on the coverslip with a needle (if it is not sealed) will often rotate the grain into the more easily recognizable view.

POLLEN STATISTICS AND THE POLLEN DIAGRAM

The pollen count

When counting pollen it is often necessary to restrict the total number of grains counted on a slide; i.e. a form of sampling is adopted rather than the counting of every pollen grain (which in some cases may exceed tens of thousands). Dimbleby (1957) did discuss the total number of pollen grains that ought to be counted to achieve a representative picture of the pollen in a sample. He found that by counting 2 samples up to 700 and 900 total pollen grains, about 20 different pollen taxa were identified, but that all taxa which have a final percentage value in excess of 1% of all pollen were identified within the first 250 grains. He did add the proviso that if one taxon is dominant then total pollen numbers would have to exceed 250 in order to ensure that all major taxa are identified. More recently some authors (e.g. Handa and Moore 1976) have adopted total pollen counts of 500 as standard and it is probable that such a level of counting would be satisfactory for most situations.

It has therefore been usual to count all pollen grains and spores until a predetermined total has been reached (i.e. the pollen sum). Traditionally a pollen sum of 150 total tree pollen (i.e. *Pinus* + *Betula* + *Quercus* + *Alnus* + *Ulmus* + *Tilia* + *Fraxinus* + *Fagus*) has been adopted for much work in north-west Europe. With a criterion like this all pollen and spores are counted until 150 tree pollen grains have been identified. This tree pollen sum has proved satisfactory when used in the analysis of material from sites that have a history dominated by tree pollen simply because pollen of shrubs and herbaceous plants together with spores provide a relatively small extra number of grains to be counted. It is less satisfactory where there is little tree pollen and this is why a total pollen sum of 500 (i.e. all pollen is counted until a total of 500 is reached) may be more useful particularly for geomorphological sites.

When counting pollen it is advisable to traverse the slide at regularly spaced intervals to avoid possible non-randomness in the distribution of pollen on the slide (see above in section on 'The microscope and pollen counting'). The traverses must be spaced sufficiently to ensure that there is no possibility of pollen being counted twice. If insufficient pollen is found while traversing one slide then a further slide (or slides) should be prepared and counted until the desired pollen total is reached.

Pollen percentages

Traditionally the figures resulting from pollen counting are expressed in simple percentage terms. Each individual pollen and spore type is expressed as a percentage of the total pollen sum. There are difficulties inherent in the use of percentages and these can be overcome by the use of absolute pollen methods. However, an adequate picture of the environment normally can be obtained through the use of simple percentages. A more detailed discussion of the statistical problems associated with pollen data can be found in, for example, Davis (1963), Mosimann (1965) and Faegri (1975).

Pollen diagrams

Once a number of levels have been counted from one profile and precentage values have been calculated, a pollen diagram can be constructed. A wide range of styles have been adopted for pollen diagrams, and while some are unsatisfactory (see Faegri 1975, for some of the most difficult to read), most of the better styles can be seen in articles in, for example, *New Phytologist, Journal of Ecology, Journal of Biogeography*. The most common style is one in which the values for individual pollen and spore types are represented in separate "saw edge" graphs (Fig 4). This style is widely accepted and easy to interpret, although the straight line between values suggests that there is a steady rise or fall in pollen values along the line when this is not necessarily the case. In order to avoid this some authors have produced "bargraph" type diagrams (e.g. Simmons 1969), where the thickness of the bars is equal to the width of sample taken from the peat profile. Sometimes these are difficult to interpret and a combination of "saw-dege" and "bargraph" has been adopted as a compromise (Hibbert, Switsur and West 1971, Pennington 1964).

THE INTERPRETATION OF POLLEN ANALYTICAL DATA

Each pollen diagram is unique as is each site and the surrounding from which the pollen rain is derived. The majority of interpretations are subjective and dependent upon the investigator's knowledge of local environments together with details of the ecology and taxonomy of the species encountered. Pollen analysis provides a figure for relative or absolute pollen frequency, not an ennumeration of particular plant species or community occurrence. It is preferable to use assemblages of pollen grains rather than individual taxa as indicators of past environments. It is possible to group taxa ecologically so that a general picture of vegetation development can be envisaged in an area, but for example it is not possible to delimit the woodland communities in a mixed deciduous forest. The major factors in the release, movement and deposition of pollen will be examined in more detail, and this will indicate the extent to which these factors influence the interpretation of pollen data.

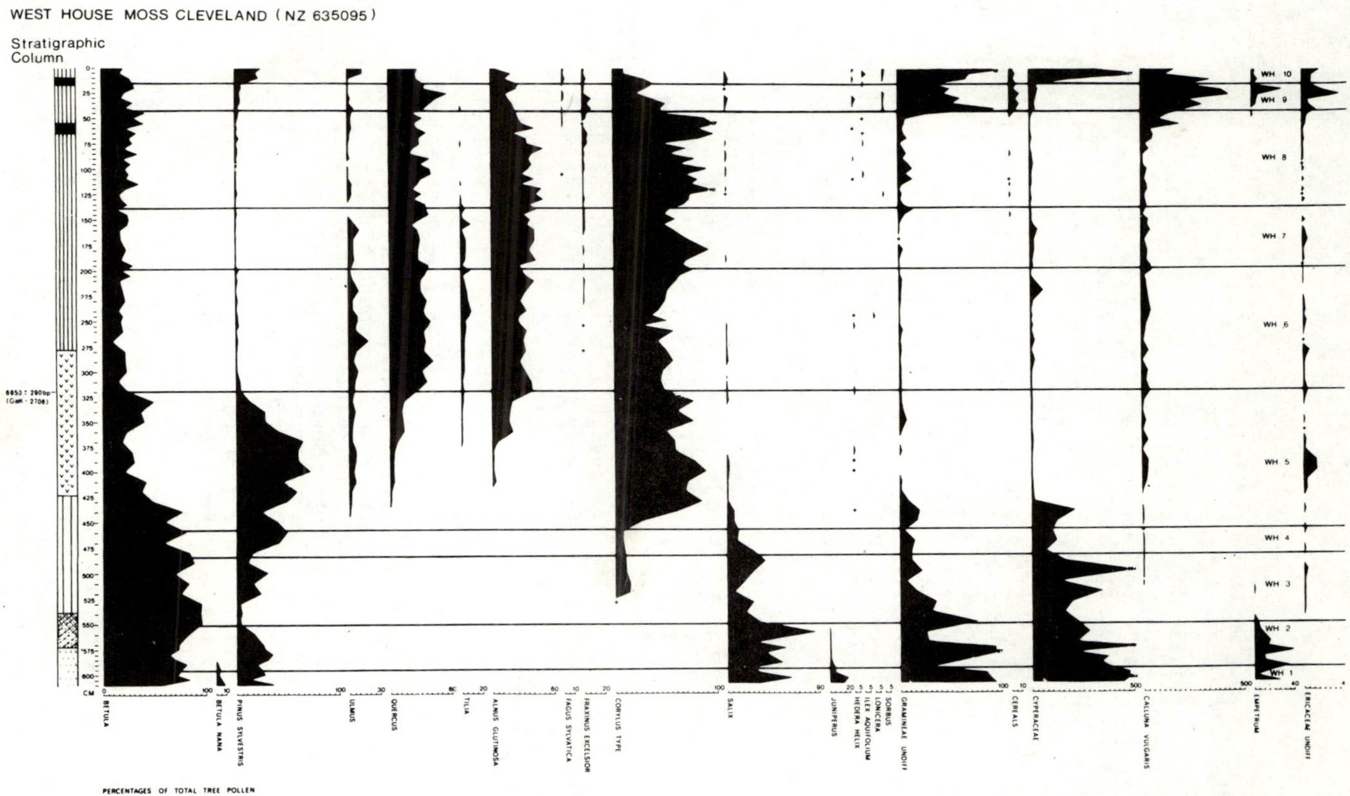

Fig. 4 Pollen diagram of 'saw edge' type with associated stratigraphic column and local pollen assemblage zones

Pollen source

It is difficult to assess the source of pollen found in deposits from a site such as West House Moss (Fig.4) although it may be acceptable to suggest that all pollen with relatively high values (i.e. at this site *Betula, Pinus, Quercus, Alnus, Corylus* type, *Salix,* Gramineae, Cyperaceae and *Calluna*) represent vegetation which was growing in close proximity to the site. When the values of the major pollen types vary from time to time this indicates that the plant communities altered over time. Low pollen values do not always indicate non-local distribution or insignificance of a plant species in the local plant community, but may reflect a low pollen production level. For example, a species such as *Hedera* is insect pollinated so the presence of *Hedera* pollen is very difficult to interpret in terms of its importance in the local plant community. Some plants such as *Calluna* might be regarded as a bog plant and therefore very locally distributed, for example at West House Moss, but if conditions were very wet in the past on this bog then *Calluna* was perhaps growing only on the drier ground around the bog. The best situation for examining the source of pollen is when the pollen spectrum in a sample can be related directly to the vegetation around the sample point. This has been carried out by several authors (e.g. Tinsley and Smith 1974, O'Sullivan 1973, Birks 1973b, Davis and Goodlet 1960) for contemporary situations. However, the usefulness of such studies for fossil interpretation is limited by the fact that in an area like north west Europe it is almost impossible to find many natural plant communities that might have been present in the past.

Pollen transport

Once pollen has been released by a plant it will travel by air and/or water or animal to a deposition site. The amount and type of pollen being transported will vary from season to season and even from day to day, and different types of site will receive pollen which has been transported by different methods (Peck 1973). Thus a closed lake will contain pollen transported by air plus that derived from terrestrial runoff, while a raised bog surface (e.g. West House Moss, Fig.4) receives most grains by air transport, and marine or estuarine sediments accumulate mostly water dispersed pollen. The nature of the different pollen taxa will influence the way in which it is transported and subsequently arrives in a particular deposit. For example, marine sediments often show high frequencies of winged conifer grains which float and hence are easily transported by water (e.g. Heusser and Balsam 1977). More detailed discussion of pollen transport, together with suggested models of transport, may be found in, for example Tauber (1965) or West (1971). The actual nature of the complex mechanisms involved in the deposition of pollen is also discussed by some of these workers. Once pollen reaches the surface of the lake, bog or soil further processes may combine to effect the final distribution of pollen. In lakes there may be differential dispersion depending upon grain density and surface area properties, and water may recirculate grains too (Davis 1968). In raised bogs (e.g. West House Moss, Fig.4) pollen may be downwashed for a short distance and differential destruction may be active. These two processes may be particularly active in soils and this means that soil pollen analyses have to be interpreted differently from results obtained from lakes and most peat bogs (Dimbleby 1961, 1967, 1977). Finally disturbance of the upper horizons of pollen profiles by animals (including man) will seriously distort the pollen record.

Stratigraphy and pollen data

The stratigraphic column shown alongside the pollen spectra in Fig.4 illustrates the major peat types sampled at West House Moss. This together with detailed examination of macrofossil remains in the peat ideally would allow a distinction to be made between the pollen from vegetation growing on the bog surface and that from vegetation around the bog. In most cases the latter would be the major vegetation type in the area. However, in practice it is difficult to determine for example whether all the Cyperaceae pollen shown in Fig.4 was derived from plants growing on the bog even if Cyperaceae fruits were found in the peat. One could suggest only that the evidence might indicate that the bulk of the Cyperaceae pollen was derived from plants on the bog. This difficulty is less significant when dealing with aquatic seeds and fruits in lake sediments where pollen of aquatic species has been found. Pollen of aquatic plants in such situations is often regarded as 'local' in origin although of course the pollen, seeds and fruits could have been washed into the lake from an inflowing stream.

Pollen zones and the pollen diagram

Where a detailed pollen diagram has been constructed it is usual practice to divide the diagram into pollen zones in order to simplify discussion of the sequence of vegetation changes which have occurred. In the past some confusion has arisen because of the indiscriminate use of stratigraphy, pollen and absolute dating in the zonation of pollen diagrams (Crabtree 1968) and equally serious has been the use of named pollen zones to signify geological or geomorphological events, for example in the term "Zone III glaciation". Sometimes stratigraphic and pollen boundaries coincide and some workers have used stratigraphic divisions in preference to pollen changes as a basis for zonation (e.g. Donner 1958). However such zonation schemes are not entirely satisfactory as changes in sedimentation in a basin may not be linked directly with vegetation changes which occur in the area around the basin. Probably the most satisfactory scheme of zonation is one which is based upon assemblages of the dominant pollen types and is named after them. The assemblage zone concept based upon the American Code of Stratigraphic Nomenclature (1963) and first employed palynologically by Cushing (1967) has recently come into widespread use in Britain. Using this system a series of local and regional zones have been formally

Stage	^{14}C Time	West Chronozones	Blytt and Sernander Periods	Godwin Zones	Vegetation Characteristics
FLANDRIAN	1000 AD BC	Fl III	Sub-Atlantic	VIII	Alnus-Quercus-Betula
	1000 2000 3000		Sub-Boreal	VIIb	Alnus-Quercus-Tilia Arboreal Pollen decline
	4000	Fl II	Atlantic	VIIa	Alnus-Quercus-Ulmus-Tilia
	5000 6000	Fl I	Boreal	VI	Pinus-Corylus: c Quercus-Ulmus-Tilia b Quercus-Ulmus a Ulmus-Corylus
	7000			V	Corylus-Betula-Pinus
	8000		Pre-Boreal	IV	Betula-Pinus Arboreal Pollen rise
LATE DEVENSIAN	9000			III	Herbs
				II	Betula-herbs
	10,000			I	Herbs

Fig. 5 Schemes of change, division and correlation in the late Devensian and Flandrian of England and Wales.

described and these are superceding the Godwin scheme (1940, and Fig.5), although this is still used by many as cross reference. West (1970) proposed chronozones (Fig.5), broad divisions which are easily subdivided. When assemblage zones are dated by radiocarbon they can be correlated with these with a good degree of synchroneity as Hibbert, Switsur and West (1971) and Hibbert and Switsur (1976) have demonstrated from Flandrian sites in Britain.

The pollen diagram from West House Moss (Fig.4) has been divided into pollen assemblage zones which have been given the local prefix WH. Thus WH1 may be described as a *Betula-Salix-Juniperus*-Gramineae-Cyperaceae zone and WH2 as a *Betula-Salix*-Gramineae-Cyperaceae zone. It can be seen that the pollen assemblage zones do not coincide with changes in stratigraphy indicating that changes in sedimentation did not occur at the same time as major vegetation changes around West House Moss. Through the use of pollen assemblage zones there is no assumption that particular vegetation changes are synchronous from site to site. This is an important principle as Birks (1973a) has demonstrated from two sites only 10 km apart on the Isle of Skye that there was a 2000 year delay in the establishment of birch forest.

The interpretation of pollen diagrams is clearly difficult and it is hoped that this outline provides a simple introduction to the basic principles of interpretation.

CONCLUSION

Pollen analysis has been developing in a number of different directions over the past few years. The introduction of numerical and computational methods in the analysis and presentation of data (e.g. Squires and Holder 1970, Deacon 1972, Gordon and Birks 1972, 1974, Birks and Deacon 1973, Birks 1974, Damblon and Schumacker 1972) is one example of this. The increase in use of absolute pollen analysis coupled with chemical and other palaeolimnological techniques (e.g. Pennington, Haworth, Bonny, and Lishman, 1972) and the detailed study of peat (e.g. Iversen 1960, Turner 1965, 1970, Dickinson 1975, Moore and Chater 1969) which span particularly interesting events indicate other directions in which the technique has already developed. For further information of these more recent developments, and more general critical reviews of pollen analysis (e.g. Crabtree 1975, Oldfield 1970), the appropriate texts should be consulted.

APPENDIX 1: BASIC ITEMS NEEDED IN A POLLEN LABORATORY

Major Equipment

- Balance
- Centrifuge
- Fume cupboard
- Hot plate
- Magnetic stirrer
- Microscope (binocular)
- Sieves
- Slide cabinet
- Still or de-ioniser
- Tally counter
- Thermometer
- Vortex mixer
- Water bath

Hardware, Glassware and Miscellaneous

- Aluminium foil
- Beakers
- Boiling tubes
- Bunsen burners
- Centrifuge tubes
- Cleaner (e.g. Teepol, Decon, Quadralene)
- Cotton wool
- Crucibles
- Dispensing bottles
- Evaporating dishes
- Eye protectors
- Filter papers
- Funnels
- Glass marking pen
- Glass phials
- Glassware draining racks
- Microscope slides and cover-slips
- Petri dishes
- Polythene bags
- Pipettes - Pasteur and Graduated 'E' Mil Gold Line
- Reagent bottles
- Retort stand bases, rods, rings and clamps
- Rubber and plastic tubing
- Rubber gloves - heavy duty
- Spatulas
- Stirring rods
- Seekers
- Slide and bottle labels
- Slide boxes
- Test tube holders
- Tongs
- Tripod stands

- Graduated cylinders
- Gas lighter
- Laboratory coats
- Test tube cleaning brushes
- Wire gauzes with asbestos centre

Chemicals

Must be of analar or above grade; technical grades often contain contaminants.

- Acetic anhydride
- Acetone
- Bromoform
- Calcium gluconate
- Ethanol
- Glacial acetic acid
- Hydrocholoric acid
- Hydrofluoric acid
- Nitric acid
- Potassium hydroxide
- Sodium carbonate
- Sodium chlorate
- Sodium hexametaphosphate
- Sodium hydroxide
- Sulphuric acid
- Toluene

Stains and mounting media

- Bismarck brown
- DPX mountant
- Fuchsin
- Glycerine jelly
- Glycerol
- Nail varnish - Elmers glue-all
- Paraffin wax
- Safranin (Y and O)
- Silicone fluid

Immersion oil

- Anisole

APPENDIX 2: LIST OF SUPPLIERS

Field Equipment Suppliers

J. H. Steward Ltd.,
Enbeeco House,
Carlton Park,
Saxmundham,
Suffolk,
IP17 2ND

(Hiller samplers and augers)

A. and H. Duncan Ltd.,
97 Crown Street,
Aberdeen,
AB9 2AQ

(peat Survey of Scotland (Russian samplers)

Leonard Farnell and Co Ltd.,
Station Road,
North Mymms,
Hatfield,
Hertfordshire.

(Hiller samplers and augers)

Abbey Tool and Gauge,
The Knowes,
Kelso,
Roxburghshire.

(Abbey (Dachnowski) sampler)

Engineering Laboratory Equipment Ltd.,
Frogmore Road,
Hemel Hempstead,
Hertfordshire,
HP3 9RL

(Augers)

Borros and Co.
Solna,
Sweden

(Hiller samplers)

Laboratory Equipment Suppliers

Griffin and George Ltd.,
285 Ealing Road,
Alperton,
Wembley,
Middlesex,
HAO 1HJ

(General items and Chemicals)

B.D.H. Chemicals Ltd.,
Poole,
Dorset,
BH12 4NN

A. Gallenkamp Ltd.,
Technico House,
Christopher Street,
London EC2.

(General items and chemicals)

Chance Ltd.,
Spon Lane,
Smethwick,
Warley.

(Microscope slides and coverslips)

Laboratory Equipment Suppliers (cont)

Baird and Tatlock,
Hopkin and Williams,
P O Box 1,
Freshwater Road,
Chadwell Heath,
Romford,
Essex,
RM1 1HA

(General items and chemicals respectively)

Allergon AB,
Valinge 55,
S26200
Engelholm,
Sweden

(Pollen type material)

Oertling Ltd.,
Cray Valley Road,
St. Mary Cray,
Orpington,
Kent,
BR5 2HA

(Balances)

Greer Laboratories
P O Box 800,
Lenoir,
N. Carolina 28645
U.S.A.

(All allergens)

M.S.E. Scientific Instruments,
Manor Royal,
Crawley,
Sussex.

(Centrifuges)

Vickers Instruments,
26 Breakfield,
Coulsden,
Surrey,
CR3 2UP

(Microscopes)

Carl Zeiss (Oberkochen) Ltd.,
Dagenhardt House,
31-36 Foley Street,
London W1P 8AP

(Microscopes)

Thomas P. Headland Ltd.,
Melon Road,
London,
SE15 5QP

(Tally Counters)

APPENDIX 3: ESTABLISHING A REFERENCE COLLECTION OF POLLEN AND SPORES

Type material can be obtained from a variety of sources. Cheapest is the field collection of flowers or the acquisition of herbarium specimens, more expensive ways include the purchase of phials of prepared, pure pollen concentrate from commercial suppliers (Appendix 2).

Reference material should be prepared in a number of ways as it is advisable to match the reference slide consulted to the method of preparation of the fossil slide. Where fresh or dried material is used, grind the flower stamens in a mortar, wash through a fine sieve with a jet of alcohol into a 15 ml centrifuge tube and centrifuge. Acetylate for 10 minutes (p 15 above) with a 1 : 1 mixture of glacial acetic acid and concentrated sulphuric acid. Follow with a second acetylysation using 9 : 1 acetic anhydride and concentrated sulphuric acid. This clears excessive debris, sharpens the structure of the pollen grains and fills them out without undue swelling (Andrew 1970). Wash and prepare material for mounting in silicone fluid or glycerine jelly stained with safranin. Mount on thin slides (1 mm): use No. 0 coverslips and seal with DPX or a similar mountant. The slide collection should be labelled appropriately and catalogued according to some system, for example, the Cambridge (Andrew 1970).

APPENDIX 4: SAMPLES FOR RADIOCARBON ASSAY

The most unsuitable looking material may contain enough carbon 14 to allow its half-life and age in radiocarbon years BP to be calculated by the process of radiometric dating. Deposits and objects of all kinds found in a variety of situations can come under consideration. The preliminaries to the submission of a sample for dating involve techniques calling for the utmost care and provision both in the field and laboratory. Circumstances may dictate that a limited amount of material is all that is available, such as the discovery of an artifact or a sliver of organic detritus in a glacial deposit. In addition, each laboratory has its own requirements, rules, methods of analysis and range of charges for accomplishing a radiocarbon assay. Reference to individual laboratories must therefore be made before collection, preparation, and submission of a sample takes place. Hence only very general guidelines can be given here as to the procedures which should be adopted. Most laboratories have an order of preference of material for dating ranging from charcoal through wood, peat and bone to shell which is least suitable. Samples for dating must be secured with extreme care whenever possible. Their provenance must be precisely delimited and the sample location should be out of range of penetration by modern roots and other contaminants involving 'young' or 'old' carbon (Shotton 1972). Where samples are removed from cores taken with a sampling device, a large diameter slice of a minimal vertical thickness (1-2 cm) is required in order to obtain a date (which inevitably will have a standard error of some magnitude) as near as possible to the point in time of depostion of the sediment. Samples should be oven dried individually and then stored in double wrapped polythene bags labelled on the outside in indelible felt pen with summary detail of that submitted formally to the laboratory. Most laboratories can produce a reliable date with *circa* 5 gm of carbon. If carbon content is high, a dry weight of 30 gm will normally provide such an amount; if it is low (for example, in organo-mineral deposits) 200-300 gm may be required to achieve the desired total. It is always better to collect more than is thought necessary, then a duplicate sample is available if required.

Radiocarbon Dating Laboratories

A selected list of major laboratories; comprehensive details are available in yearly issue 3 of *Radiocarbon*. This Journal should be consulted for further information concerning assays.

Code

A ARIZONA
Dr. Austin Long, Laboratory of Isotope Geochemistry, Geosciences Department,
University of Arizona, Tucson, Arizona 85721.

Code

ANU AUSTRALIAN NATIONAL UNIVERSITY
Mr. H.A Polach, ANU Radiocarbon Dating Laboratory, c/o Rsearch School of Earth Sciences, The Australian National University, Canberra ACT, Australia 2600.

BIRM BIRMINGHAM
R.E.G. Williams, Department of Geological Sciences, P O Box 363, University of Birmingham, Birmingham. B15 2TT

BM BRITISH MUSEUM
Mr. Richard Burleigh, Research Laboratory, The British Museum, London WC1B 3DG

Gak GAKUSHIN UNIVERSITY
Prof. K. Kigoshi, Gakushin University, Mejiro, Toshima ku, Tokyo, Japan.

GrN GRONINGEN
Dr. W.G. Mook, Natuurkundig Laboratorium der Rijks Universiteit, Westersingek 34, Groningen, Netherlands.

HAR HARWELL
R.L. Otlet, Carbon 14/Tritium Measurements Laboratory, Building 10-46 AERE, Harwell, Didcot, Berkshire OX11 ORA

I TELEDYNE ISOTOPES
James Bücklet, Teledyne Isotopes, 50 van Buren Avenue, Westwood, New Jersey 07675

K COPENHAGEN
Dr. Henrick Tauber, Geological Survey of Denmark and National Museum, DK-1220 Copenhagen K. Denmark.

Lu LUND
Prof. Björn Berglund, and Mr. Sören Håkansson, Radiocarbon Dating Laboratory, University of Lund, Tunavagen 29, S-22363 Lund, Sweden.

NPL NATIONAL PHYSICAL LABORATORY
Mr. W.J. Callow and Miss G.I. Hassal, Division of Radiation Science, National Physical Laboratory, Teddington, Middlesex

Q CAMBRIDGE
Prof. R.G. West or Dr. V.R. Switsur, University Sub-Department of Quaternary Research, Botany School, Downing street, Cambridge.

Code	
QL	QUATERNARY ISOTOPE LABORATORY Minze Stiuver, Quaternary Research Center, University of Washington, AK-60 Seattle, Washington 98195
SRR	SCOTTISH UNIVERSITIES RESEARCH AND REACTOR CENTRE Dr. D.D. Harkness and Prof. H.W. Wilson, Scottish Universities Research and Reactor Centre, East Kilbride, Glasgow G75 0QU
ST	STOCKHOLM Dr. Eric Welin, Laboratory for Isotope Geology, Swedish Museum of Natural History, S 104 05, Stockholm 50, Sweden.
T	TRONDHEIM Dr. Reider Nydal, Mr. Steiner Gulliksen and Mr. Knut Löuseth, Radiological Dating Laboratory, The Norwegian Institute of Technology, Trondheim, Norway.
U	UPPSALA Dr. Ingrid U. Olsson, Institute of Physics, University of Uppsala, Box 530, S 751 21, Uppsala, Sweden.
UB	BELFAST Mr G.W. Pearson or Dr. J.R. Pilcher, Palaeoecology Laboratory, The Queen's University, Belfast, BT7 1NN
UCLA	UNIVERSITY OF CALIFORNIA, LOS ANGELES Dr. Rainer Berger and Dr. W.F. Libby, Institute of Geophysics, University of California, Los Angeles California.
W	U S GEOLOGICAL SURVEY Dr. Meyer Rubin, U.S. Geological Survey National Center, 971, Reston, Virginia, 22092
WIS	WISCONSIN Dr. Margaret Bender, Radiocarbon Laboratory of the Center for Climatic Research, Institute for Environmental Studies, University of Wisconsin 1225 W Dayton St., Madison, Wisconsin, 53706

REFERENCES

American Commission on Stratigraphic Nomenclature (1961) Code of Stratigraphic Nomenclature. *Bull. Am. Assoc. Petr. Geol.* 45, (5), 645-665.

Andersen S.T. (1965) Pollen analysis and till stratigraphy at Lindø, Denmark. pp 65-78 in *International Studies on the Quaternary* (H.E. Wright Jr. and David G. Frey eds). GSA Special Paper 84, New York.

Andrew R. (1970) The Cambridge pollen reference collection. pp 225-231 in *Studies in the Vegetational History of the British Isles* (D. Walker and R.G. West eds.). Cambridge University Press.

Bakker van Zinderen E.M. (1964) A pollen diagram from equatorial Africa: Cherangani, Kenya. *Geologie Mijnb* 43, 123-128.

Barber K.E. (1976) History of vegetation. pp 5-83 in *Methods in Plant Ecology* (S.B. Chapman ed.). Blackwell Scientific Publications, Oxford.

Bell F.G. (1970) Late Pleistocene floras from Earith, Huntingdonshire *Phil. Trans. R. Soc.* B 258, 347-378.

Beninghoff W.S. (1962) Calculation of pollen and spore density in sediments by addition of exotic pollen in known quantities (Abstract). *Pollen Spores* 4, (2) 332 - 333.

Birks H.J.B. (1973a) *Past and Present Vegetation of the Isle of Skye. A Palaeoecological Study.* Cambridge University Press.

Birks H.J.B. (1973b) Modern pollen rain studies in some Arctic and Alpine environments. pp 143-168 in *Quaternary Plant Ecology* (H.J.B. Birks and R.G. West eds.). Blackwell Scientific Publication, Oxford.

Birks H.J.B. (1974) Numerical zonation of Flandrian pollen data. *New Phytol.* 73, 351-358.

Birks H.J.B. and Deacon J. (1973) A numerical analysis of the past and present flora of the British Isles. *New Phytol.* 72, 877-902.

Blackburn K.B. (1931) The late-glacial and post-glacial periods in the North Pennines. II. Possible survivals in our flora. *Trans. North. Nat. Union* 1, 30-36.

Bonny A.P. (1972) A method for determining absolute pollen frequencies in lake sediments. *New Phytol.* 71, 393-405.

Brookes D and Thomas K.W. (1967) The distribution of pollen grains on microscope slides. Part 1. The non-randomness of the distribution. *Pollen spores* 9, 621-630.

Brookes C.L. (1972) Pollen analysis and Main Buried Beach in the western part of the Forth Valley. *Trans. Inst. Br. Geogr.* 55, 161-170.

Brown C.A. (1960) *Palynological Techniques.* Privately published, Baton Rouge.

Chapman S.B. (ed) (1976) *Methods in Plant Ecology.* Blackwell Scientific Publications, Oxford.

Clymo R.S. (1964) Apparatus for recovery of lost peat borers. *New Phytol.* 63, 426-427.

Crabtree K. (1968) Pollen analysis. *Sci. Prog. Oxf.* 56, 83-101.

Crabtree K. (1975) A review of methodological advances in pollen analysis. pp 266-283 in *Processes in Physical and Human Geography* (R. Peel, M. Chisholm and P. Haggett eds.). Heinemann Educational Books Ltd., London.

Cushing E.J. (1967) Late Wisconsin pollen stratigraphy and the glacial sequence in Minnesota. pp 59-88 in *Quaternary Paleoecology* (E.J. Cushing and H.E. Wright eds.). Yale University Press, New Haven and London.

Damblon F. and Schumacker R. (1972) New prospects for the study of palynological data: the use of computers. *Pollen Spores* 13, 609-614.

Davis M.B. (1963) On the theory of pollen analysis. *Am.J. Sci.* 261, 897-912.

Davis M.B. (1965) A method for determination of absolute pollen frequency. pp 674-686 in *Handbook of Paleontological Techniques* (B. Kummell and D. Raup eds.) Freeman, San Francisco.

Davis M.B. (1967) Pollen Accumulation rates at Rogers Lake, Connecticut during Late - and Post-glacial time. *Rev. Palaeobotan. Palynol.* 2, 219-230.

Davis M.B. (1968) Pollen grains in lake sediments: redeposition caused by seasonal water circulation. *Science N.Y.* 162, 796-799.

Davis M.B. and Goodlet J.C. (1960) Comparison of the present vegetation with pollen spectra from Browningtom Pond, Vermont. *Ecology* 41, 804-825.

Deacon J. (1972) A data bank of Quaternary plant fossil records. *New Phytol.* 71, 1337-1232.

Dickinson W. (1975) Recurrence surfaces in Rusland Moss, Cumbria (formerly north Lancashire). *J. Ecol.* 63,(3), 913-935.

Dimbleby G.W. (1957) Pollen analysis of terrestrial soils. *New Phytol.* 56, 12-28.

Dimbleby G.W. (1961) Soil Pollen Analysis. *J. Soil Sci.* 12, 1, 1-11.

Dimbleby G.W. (1967) *Plants and Archaeology* John Baker, London

Dimbleby G.W. (1977) *Ecology and Archaeology* Edward Arnold, London.

Donner J.J. (1958) Loch Mahaick, a late glacial site in Perthshire. *New Phytol.* 57, 183-186.

Erdtman G. (1928) Studies in the post-artic history of the forests of the North-west Europe I. Investigations in the British Isles. *Geol. Foren. Stock. Forh.* 50, 123-192.

Erdtman G. (1929) Some aspects of the post-glacial history of British forests. *J. Ecol.* 17, 112-126.

Erdtman G. (1943) *An Introduction to Pollen Analysis.* Chronica Botanica, Waltham, Mass.

Erdtman G. (1969) *Handbook of Palynology.* Munksgaard, Copenhagen.

Erdtman G. Berglund B. & Praglowski J. (1961) An introduction of a Scandinavian Pollen Flora. *Grana Palynologica* 2, 3-92.

Evans G.H. (1970) Pollen and diatom analyses of late-Quaternary deposits in the Blelham basin, north Lancashire. *New Phytol.* 69, 821-874.

Faegri K. (1975) *Textbook of Pollen Analysis* (3rd Edn). Munksgaard, Copenhagen and Blackwell Scientific Publications Oxford.

Franks J.W. & Johnson R.H. (1964) Pollen analytical dating of a Derbyshire landslip: the Cown Edge landslides, Charlesworth. *New Phytol*. 63, 209-216.

Fredskild B. (1969) A postglacial standard pollen diagram from Peary Land, North Greenland. *Pollen Spores*. 11, 573-583.

Godwin H. (1934) Pollen analysis: an outline of the problems and potentialities of the method. Part I. Techniques and interpretation. Part II. General applications of pollen analysis. *New Phytol*. 33, 278-305 (I). 325-358 (II).

Godwin H. (1940) Pollen analysis and forest history of England and Wales. *New Phytol*. 39, 370-400.

Godwin H. (1943) Coastal peat beds of the British Isles and North Sea. *J. Ecol*. 31, 199-247.

Godwin H. (1975) *The History of the British Flora* (2nd Edn.), Cambridge University Press.

Gordon A.D. and Birks H.J.B. (1972) Numerical methods in Quaternary palaeoecology. I. Zonation of Pollen diagrams. *New Phytol*. 71, 961-979.

Gordon A.D. and Birks H.J.B. (1974) Numerical methods in Quaternary palaeoecology. II. Comparison of pollen diagrams. *New Phytol*. 73, 221-249.

Gregory P.H. (1973) *The Microbiology of the Atmosphere* (2nd Edn.), Leonard Hill, London.

Handa S. and Moore P.D. (1976) Studies in the vegetational history of mid Wales. IV. Pollen analyses of some pingo basins. *New Phytol*. 75, 205-225.

Haworth E.Y. (1969) The Diatoms from a sediment core from Blae Tarn, Langdale. *J. Ecol*. 57, 429-441.

Heusser L. and Balsam W.L. (1977) Pollen distribution in the north-east Pacific Ocean. *Quaternary Research*. 7,(1), 45-62.

Hibbert F.A. Switsur V.R. & West R.G. (1971) Radiocarbon dating of Flandrian pollen zones at Red Moss, Lancashire. *Proc. Roy. Soc*. B 177, 161-176.

Hibbert F.A. and Switsur V.R. (1976) Radiocarbon dating of Flandrian pollen zones in Wales and Northern England. *New Phytol.* 77, 793-807.

Hope G.S. (1976) The vegetational history of Mt. Wilhelm, Papua New Guinea. *J. Ecol.* 64,(2), 627-663.

Hyde H.A. and Adams K.F. (1958) *An Atlas of Airborne Pollen Grains.* Macmillan, London.

Iversen J. (1960) Problems of the early post-glacial forest development in Denmark. *Danm. Geol. Unders.* IV, 4,(3),1-32.

Jowsey P.C. (1966) An improved peat sampler. *New Phytol.* 65, 245-248.

Kummell B. and Raup D. (eds.) (1965) *Handbook of Paleontological Techniques.* Freeman, San Francisco.

Mackereth F.J.H. (1958) A portable core sampler for lake deposits. *Limnol Oceanogr.* 3, 181-191.

Mackereth F.J.H. (1966) Some chemical observations on post-glacial lake sediments. *Phil. Trans Roy. Soc. Lond.* B, 250, 165-213.

Matthews J. (1969) The assessment of a method for the determination of absolute pollen frequencies. *New Phytol.* 68, 161-166.

Moore P.D. and Chater E.H. (1969) The changing vegetation of west-central Wales in the light of human History. *J. Ecol.* 57, (2), 361-379.

Mosimann J.E. (1965) Statistical methods for the pollen analyst. Multinomial and negative multinomial techniques. pp 636-673 in *Handbook of Paleontological Techniques* (B. Kummel and D. Raup, eds.), Freeman, San Francisco.

Newey W.W. (1966) Pollen analysis of sub-carse peats of the Forth Valley. *Trans. Inst. Br. Geogr.* 39, 53-59.

Oldfield F. (1970) Some aspects of scale and complexity in pollen analytically based palaeocology. *Pollen Spores.* 12, 2, 163-171.

O'Sullivan P.E. (1973) Contemporary pollen studies in a native Scots pine ecosystem. *Oikos.* 24, 143-150.

Paepe R. (1971) Dating and position of fossil soils in the Belgian Pleistocene Stratigraphy. pp 261-269 in *Paleopedology: Origin, Nature and Dating of Paleosols.* (Dan H. Yaalon ed.). International Society of Soil Science and Israel Universities Press, Jerusalem.

Peck R.M. (1973) Pollen budget studies in a small Yorkshire catchment. pp 43-60 in *Quaternary Plant Ecology* (H.J.B. Birks and R.G. West eds.). Blackwell Scientific Publications, Oxford.

Peck R.M. (1974) A comparison of four absolute pollen preparation techniques. *New Phytol.* 73, 567-587.

Pennington W. (1964) Pollen analyses from the deposits of six upland tarns in the Lake District. *Phil. Trans R. Soc.* B 248, 205-244.

Pennington W. (1973) Absolute pollen frequencies in the sediments of lakes of different morphometries. pp 79-104 in *Quaternary Plant Ecology* (H.J.B. Birks and R.G. West eds.). Blackwell Scientific Publications, Oxford.

Pennington W. and Bonny A.P. (1970) An absolute pollen diagram from the British late-glacial. *Nature Lond.* 226, 871-873.

Pennington W., Haworth E.Y., Bonny A.P. and Lishman J.P. (1972) Lake sediments in Northern Scotland. *Phil. Trans. R. Soc.* B 264, 191-294.

Shotton F.W. (1972) An example of hard water error in radiocarbon dating of vegetable matter. *Nature Lond.* 240, 460-461.

Simmons I.G. (1969) Pollen diagrams from the North York Moors. *New Phytol.* 68, 807-827.

Simmons I.G. and Cundill P.R. (1974) Late Quaternary vegetational history of the North York Moors. II Pollen analyses of landslip bogs. *J. Biogeogr.* 1, 253-261.

Sparks B.W. and West R.G. (1972) *The Ice Age in Britain.* Methuen, London.

Squires R.H. and Holder A.P. (1970) The use of computers on the presentation of pollen data. *New Phytol.* 69, 875-883.

Stockmarr J. (1972) Tablets with spores used in absolute pollen analysis. *Pollen Spores.* 13, 615-621.

Tauber H. (1965) Differential pollen dispersal and the interpretation of pollen diagrams. With a contribution to the interpretation of the elm fall. *Damn Geol. Unders.* (Ser II) 89, 7-64.

Thomas K.W. (1964) A new design for a peat sampler. *New Phytol.* 63, 422-425.

Tinsley H. and Smith R.T. (1974) Surface pollen studies across a woodland/heath transition and their application to the interpretation of pollen diagrams. *New Phytol.* 73, 547-565.

Tooley M.J. (1976) Flandrian sea-level changes in West Lancashire and their implications for the Hillhouse coastline. *Geol. J.* 11, 137-152.

Turner C. (1970) The Middle Pleistocene desposits at Marks Tey, Essex. *Phil. Trans. Roy. Soc.* B 257, 373-437.

Turner J. (1965) A contribution to the history of forest clearance. *Proc. Roy. Soc.* B 161, 343-354.

Turner J. (1970) Post-Neolithic disturbance of British vegetation. pp 97-116 in *Studies in the Vegetational History of the British Isles* (D. Walker and R.G. West eds.). Cambridge University Press.

van der Hammen T. and Gonzalez E. (1965) A pollen diagram from "Laguna de la Herrera" (Sabana de Bogota). *Leid. Geol. Meded.* 32, 183-191.

von Post L. (1918) Skogstradpollen i sydsvenska torvmosslagerfolder. *Forh 16 skand natureforskermϕte 1916,* 433-465 (translated in *Rev. palaeobot. palynol.* 2, 1968.

Waddington J.C.B. (1969) A stratigraphic record of the pollen influx to a lake in the Big Woods of Minnesota. *Spec. Pap. Geol. Soc. Am.* 123, 263-282.

Walker M.J.C. and Lowe J.J. (1976) The Abbey Piston Corer. *Quaternary Newsletter.* 19, 1-3.

Watts W.A. (1959) Interglacial deposits at Kilbeg and Newtown Co. Waterford. *Proc. R. Ir. Acad.* B. 60, 79-134.

Watts W.A. and Winter T.C. (1966) Plant macrofossils from Kirchner Marsh, Minnesota a palaeocological study. *Bull geol. Soc. Am.* 77, 1339-1360.

West R.G. (1957) Interglacial deposits at Bobbitshole, Ipswich *Phil. Trans. Roy. Soc.* B241, 1-31.

West R.G. (1968) *Pleistocene Geology and Biology*. Longmans, London.

West R.G. (1970) Pollen zones in the Pleistocene of Great Britain and their correlation. *New Phytol*. 69, 1179-1183.

West R.G. (1971) *Studying the Past by Pollen Analysis*. Oxford Biology Reader No. 10, Oxford University Press.